GERMAN GLIDERS
IN WORLD WAR II

DFS 230 • DFS 331 • Go 242 • Go 345 • Ka 430 • Me 321 • Ju 322

Heinz J. Nowarra

SCHIFFER MILITARY HISTORY
West Chester, PA

Bibliography:
Nowarra-Kens, *Die deutschen Flugzeuge 1933-45*
Nowarra, *Luftkrieg im Westen*
Reitsch, *Fliegen — mein Leben*
Thetford-Riding, *Aircraft of the Fighting Powers, Vol. 4*

Translated from the German by Dr. Ed Force.

Copyright © 1991 by Schiffer Publishing Ltd.
Library of Congress Catalog Number: 91-62747.

All rights reserved. No part of this work may be reproduced or used in any forms or by any means—graphic, electronic or mechanical, including photocopying or information storage and retrieval systems—without written permission from the copyright holder.

Printed in the United States of America.
ISBN: 0-88740-358-1

This title was originally published under the title,
Deutsche Lastensegler,
by Podzun-Pallas Verlag, Friedberg.

We are interested in hearing from authors with book ideas on related topics. We are also looking for good photographs in the military history area. We will copy your photos and credit you should your materials be used in a future Schiffer project.

Published by Schiffer Publishing, Ltd.
1469 Morstein Road
West Chester, Pennsylvania 19380
Please write for a free catalog.
This book may be purchased from the publisher.
Please include $2.00 postage.
Try your bookstore first.

Oberleutnant Witzig (in cap), Leader of the "Granite" Assault Group

DFS 230

At 4:30 A.M. on May 10, 1940, three Ju 52's took off from the Cologne airfields of Ostheim and Butzweilerhof, towing behind them three other aircraft that lacked motors. The first front service of a new weapon, the freight glider, had begun. In all, 41 freight gliders were launched for this mission. In each of them sat 14 paratroopers. The first group, whose training had taken place secretly under the code word "Granite", were under the command of Oberleutnant Witzig. He commanded 85 men, who were armed with handguns and 500-kilogram explosives, mostly hollow and stick grenades. This first group utilized seven DFS 230 freight gliders. It was their task to take the Belgian Fort Eben Emael near Maastrich and hold it until the first army units, a group of Engineer Battalion No. 51, arrived. Thus the crossing of the Albert Canal was to be secured. The second group, "Beton" (Concrete), under Leutnant Schacht, consisted of eleven DFS 230 freight gliders and 96 men. The third group, "Stahl" (Steel), with nine DFS 230 and 92 men under Oberleutnant Altmann, was augmented by the fourth group, "Eisen" (Iron), under Leutnant Schächter, with ten DFS 230 and 90 men. The most successful group was that of Oberleutnant Witzig, though it landed directly on the fort without its leader. The Oberleutnant's glider had to release prematurely and landed near Düren. Witzig was able, though, to obtain a new Ju 52 and fly on behind his group. He landed in the midst of Belgian artillery fire, for the battle was in full swing. It was already 8:30 A.M.

Above: An unhooked DFS 230 over Belgium.

Right: One of the armored turrets of Eben Emael that was put out of action by the Witzig group.

Only on the next morning was a battle group of engineers able to advance to meet Witzig's group. With that, the cornerstone of Belgian defense was in German hands.

How did this new weapon come to be developed?

The development of towed aircraft goes back a long way. Fokker, then an aircraft pioneer from Holland working in Germany, received a patent as early as 1912 for an airplane with a trailer. But he never made use of this patent. The idea was not reborn until fifteen years later. The glider specialist Gottlob Espenlaub put a unit in the air that consisted of a light Espenlaub E 12 (35-hp Anzani engine) sport plane and a towed E 5 glider. Espenlaub did not follow up the subject, as the empennage had blown off on take-off. That was in March of 1927.

Then the Raab-Katzenstein firm in Kassel took up the idea. Within three weeks, the RK 7 towed craft was built. On April 13, 1927 the first successful towed flight took place. The towplane was an RK 6 biplane, a copy of an old LVG biplane of World War I. The test flights were so successful that Russians, Italians and Americans soon appeared and showed an interest in the military use of towed aircraft. The Americans even bought the whole unit.

Further tests were not made in Germany for the time being. Only in the early thirties did the German Research Institute of Gliding (DFS) plan to build a large glider as a flying observatory for meteorological purposes. The idea was given up in favor of a freight glider. This was to be towed by transport planes and unhooked over towns that were not included in regular airline service. By 1933 the planning had gone so far that a test glider was built and towed by a Ju 52. The glider's pilot was Hanna Reitsch.

Now the Luftwaffe, which was still being built up, began to take notice of this development. The DFS was contracted with to

Above: The world's first tow unit, developed by Gottlob Espenlaub in 1927 (E 12 and E 5).
Below: The towplane of the first successful tow unit, Raab-Katzenstein RK 6 (a copy of the LVG B III).CAPS P5

develop a ten-seat freight glider that could drop equipment and air-land troops behind the enemy front. The well-known glider designer Hans Jacobs created this aircraft, which was given the type designation of DFS 230. When Hanna Reitsch displayed DFS 230 V-1 (first test plane), under strict secrecy, to a large group of the military leadership, including Udet, Kesselring, Greim, Model, Keller and Milch, the value of this new weapon was recognized. A second test, involving both paratroopers and air-land troops, clearly showed the advantages of the freight glider. The Reich Air Ministry immediately issued a contract for series production.

Upper left: The RD 7 "Schmetterling" (Butterfly), built and successfully tested by the firm of Raab-Katzenstein in Kassel.

Left: The generals of the Army and Luftwaffe were impressed by the performance of the DFS 230. This photo was taken after the demonstration. From left to right: Air General Kühl, General Kesselring, Hanna Reitsch, unknown, Generalleutnant Keller, unknown, Generalleutnant Model, unknown.

One of the first DFS 230 A-1 gliders on a test flight.

Above: The fuselage of DFS 230 V-1. The wing attachments and entry hatch are easy to see. Below: The wing of the first DFS 230 was considerably more rounded at the end than that of the later production model.

Left: The bolts on the entry hatch were as simple as possible, to allow a fast exit from the craft.

Right: The DFS 230 had a simple wooden empennage with linen covering.

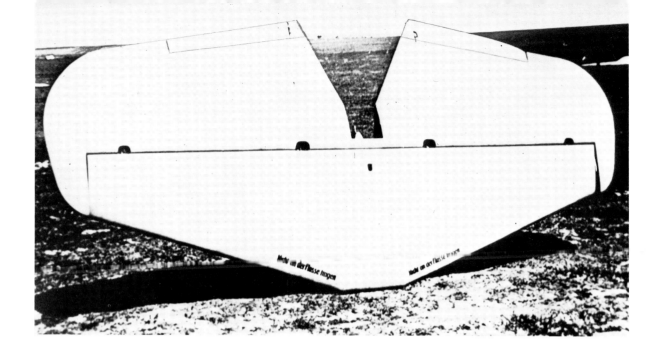

Right: Unlike the asymmetrical rudder, the elevators were symmetrical.

Left: The nose of the fuselage without its covering clearly shows the arrangement of the cockpit.

Upper left: The cockpit of the DFS 230, with instrument panel and control stick.

Above: The cockpit was separated from the crew space. The eight to ten men sat astride, one behind the other.

Left: The fuselage was constructed as a welded steel-tube frame. The seat for the crew is easy to see.

This picture shows that the freight glider lifted off first on take-off. Only when it was in the air did the towplane, here a Do 17 E, also take off.

Even fighter planes could be used to tow freight gliders. Here they are Czech Avia BH 534's, which were taken over by the Luftwaffe when Czechoslovakia was occupied. They were still used by a fighter unit in 1939, then turned over to flying schools.

During the war, more and larger freight gliders were developed, not only by the DFS but also by aircraft manufacturers, but the DFS 230 still saw service in various ways in 1944. On September 12, 1943 Mussolini, who was held prisoner at Campo Imperatore on the Grand Sasso in the Abruzzi, was rescued by members of the 1st Company of the Paratroop Instructional Battalion, under the command of Oberleutnant von Berlepsch, who landed their DFS 230 right next to the hotel in which Mussolini was being held. In the spring of 1944, a mission was flown with DFS 230's against Tito's headquarters near Drvar in Bosnia, but without success.

Next came the B-1, equipped with brakes to shorten landings. The C-1 version, planned especially for the Mussolini undertaking, was made by rebuilding the A-1. It had three braking rockets in the nose. In addition, this version also had its entry hatch ahead of the wing strut. The DFS 230 V-7 appeared in 1944, but this was developed by Chief Designer Hühnerjäger at the Gotha Wagon Factory. This craft, which was to go into production as DFS 230 F-1, had fixed landing gear with a tail skid, unlike the droppable landing gear of the normal DFS 230. It was considerably roomier and carried 15 fully equipped soldiers plus a two-man crew.

Various models were used as towplanes for the DFS 230. In addition to the Ju 52 used at first, the Henschel Hs 126, Stuka Ju 87, Dornier Do 17 E and Do 17 Z were used later, as well as the former Czech Avia BH 534 fighter.

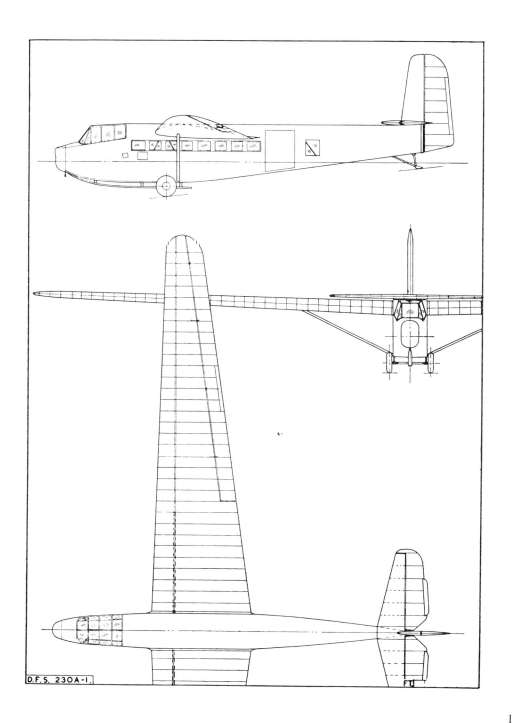

Left: The paint of all freight gliders was changed frequently. The changes were generally carried out by the troops themselves to suit the combat area. This DFS 230 shows a paint job rarely used for gliders: it is painted in dark green and greenish-black segments above and sky blue below, quite unlike the LC+1-186 on page 6.

Below: The shortage of planes, especially of Ju 52's, made it necessary to use dive bombers as towplanes for gliders. The DFS 230 proved itself here too, though not in offensive tasks, but rather in supplying advanced or cut-off units. Here a Ju 87 R-2 is seen towing a DFS 230.

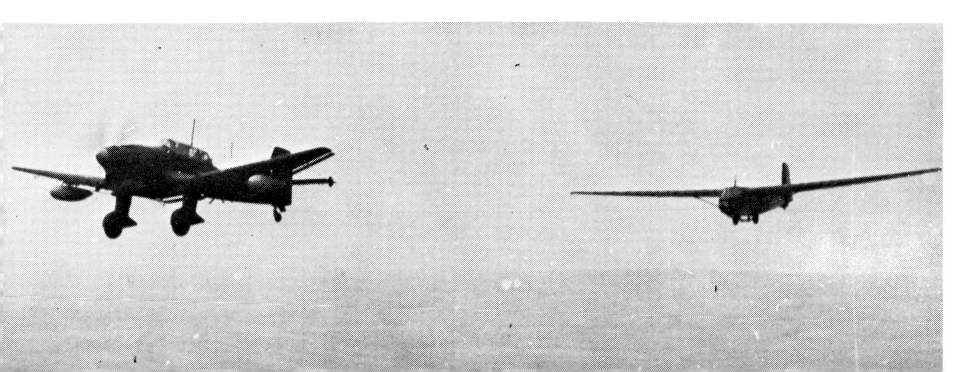

This is how the radioman aboard a Ju 87 saw the glider being towed by his own plane, as well as the glider on tow beside them. For lack of fighter protection, the towplanes also had to provide defense against enemy fighters.

Non-flexible Towing

The introduction of non-flexible towing was preceded by intensive testing. At the top and at left, the same glider is being towed first by an H 111 H-6 (upper left) and then by a Ju 52. Above is another photo of a non-flexible tow by a Ju 52. By the swastika on a white background and the incomplete radio numbers on the fuselage and wings of the Ju 52 it can be seen that these pictures were taken during testing.

Makeshift Towplanes

Above and left: The DFS 230 saw action in the Mediterranean and Balkan areas too. The Henschel HS 126 short-range reconnaissance plane fulfilled its task as a towplane just as well as did the Ju 52 in 1940. Action was also seen by the Ju 87, Do 17 and the ex-Czech Avia BH 534 fighter.

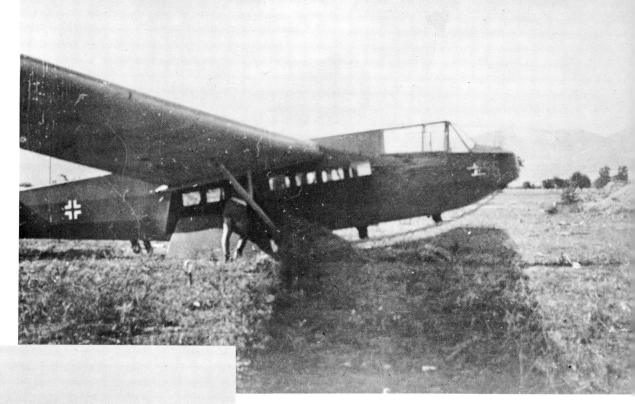

Right: In the mountainous regions of the Balkans, the light gray spotted camouflage paint was not used on the DFS. Here the old dark green and greenish-black paint was preferred.

Left: At Whitsun in 1944, an attempt was made to take the Yugoslav partisan leader Josip Broz-Tito by surprise at his headquarters in Drvar, Bosnia. DFS 230 gliders were used here too, but without success.

Left: The so-called "Mistletoe" unit was developed by the German Research Institute of Gliding, under the direction of Fritz Stamer, using a Messerschmitt Bf 109 and a DFS 230 A-1.

Right: Since the view downward from the DFS 230 never completely corresponded to combat requirements, the full-view canopy shown here was tested, but never went into production.

The wish to widen the too-narrow fuselage, the low towing speed, and the not always pleasant flying characteristics of the DFS 230 led to the building of an improved test craft, the DFS 230 V-7.

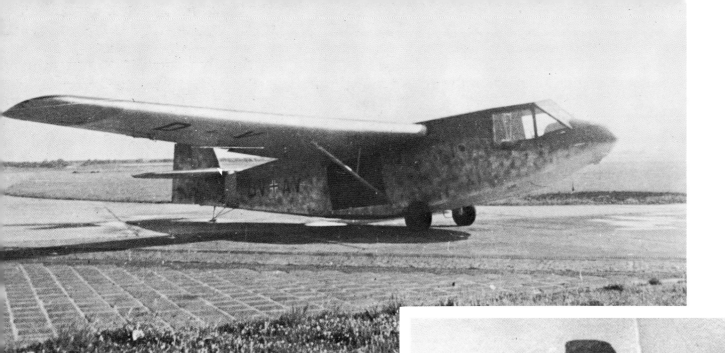

Left: The DFS 230 V-7, which was to be the model for the planned F-1 series, had fixed landing gear with a tail skid. The wingspan was shorter, but the surface was enlarged.

Right: In addition to its side entry hatch, DFS 230 V-7 also had a big loading hatch in the fuselage roof, which made it possible to load bulky articles.

Between 1938 and 1943, the DFS developed eleven different towing procedures:

1. Short tow: Towline only 1 to 1.5 meters long. Not readied for troop use.
2. Non-flexible tow: Rejected because of too-great weight on the non-flexible connector.
3. Long and non-flexible tow: Tested 1940-41. Towline could be pulled in from the towplane in flight, if necessary.
4. New non-flexible towing device made by Weserflugzeugbau worked very well, so that its introduction for troop use was ordered.
5. Multiple tow: Large glider towed by 2 or 3 towplanes. Development began in 1938 but was only carried out with Me 321 in 1940-41.
6. Towing away: Damaged craft were removed by several other planes via coupling in flight. Not introduced.
7. Carry towing: Only experiments.
8. Catch towing: According to a Hungarian patent, cables from towplanes should be caught during overflight and gliders towed from a standstill. Proved to be impossible in practice.
9. Mistletoe towing: Motorized plane is firmly attached to the glider and lifts it by its engine power. Was later used with Ju 88 A-4 and G-1 planes in mistletoe attacks. Tests with DFS 230 freight gliders were made with Klemm 35, Focke-Wulf 56 and Bf 109 E planes.
10. Carried take-off: The idea was that a towed glider could carry part of the weight of an overloaded bomber or transport at take-off. Carried out with Do 17 MV-1 and DFS 230. Made superfluous by rocket devices.
11. Volga take-off: Take-off assistance for overloaded powered planes and freight gliders with the help of ground vehicles. Tested with DFS 230 linked to personnel carrier (by chain) or two locomotives (!). Tests broken off.

An interesting development that was said to have proved useless after testing was the Focke-Achgelis Fa 225 tow-copter. This was a DFS 230 A-1 which, instead of wings, had the rotor of a Focke-Achgelis Fa 223 helicopter. Various flights were made with this craft, towed by a Ju 52. But since no advantages over a normal freight glider resulted, the development was given up.

Below: The only defensive armament of the DFS 230 A-1 was an MG 15 machine gun.

Left: the Focke-Achgelis Fa 225 tow-copter was a DFS 230 with the rotor of an Fa 223. This rotor had a diameter of 11.7 meters. The entire craft weighed only 2000 kilograms.

Right: The only Fa 225 built was used along with normal DFS 230 A-1 gliders on September 12, 1943, in the rescue of Mussolini at the Gran Sasso massif in the Abruzzi. The undertaking was carried out by Otto Skorzeny and the 1st Company of the Paratroop Training Battalion under Oberleutnant von Berlepsch. The picture shows the Fa 225 being towed by a Ju 52.

DFS 331

Around 1942 the DFS 331 heavy freight glider was created in collaboration with Gotha, but only one example was built. It was characterized by a very glazed nose with raised pilot's seat covering, double rudders, and a rear fuselage section that folded up behind the wings. It was intended to transport bulky goods, such as light Flak guns, VW personnel carriers and similar articles.

Above: The only DFS 331 V-1 had a particularly wide and roomy fuselage. This could fold upward behind the trailing edge of the wings for loading. Its landing gear consisted of two wheels that were dropped after take-off.

Left: Front view of the DFS 331 V-1, showing the unusual width of the fuselage and the good panoramic view for the pilot, who sat on the port side.

Go 242 A-1

The Gotha Go 242 A-1 freight glider was planned from the start by its designer, Dipl.Ing. Kalkert, so that a relatively inexpensive and practical transport plane could be made of it by extending the two empennage carriers forward and installing engines.

The cooperation of the Gotha Wagon Factory and the DFS in the development of the DFS 230 has already been mentioned. Gotha's new chief designer, Dipl.Ing. A. Kalkert, was given a contract in 1941 to develop a glider to carry 20 fully equipped soldiers or a corresponding load. Kalkert followed new trails in design. He decided in favor of a high-wing craft with two empennage booms, allowing a large cargo space with a rear fuselage that folded up. The new model, designated Go 242, could be flown on a cable or non-flexible link.

The flying characteristics were better than those of the DFS 230. When the first Heinkel He 111 Z (twins) were available in 1942, these towplanes could tow two or three Go 242's. In the winter of 1941-42 the Go 242 was also tested with snow skids, first at Rechlin at the Luftwaffe test center, then at Dorpat, Estonia. The towplane was an He 111 H-6. In normal snow conditions, and with a load of 2500 kg, the Go 242 had a take-off weight of 12,200 kg. Despite this comparatively high weight, the Go 242 lifted off behind an He 111 H-6 after just 1000 meters. The landing distance was only 250 meters, which was of particular value on field airstrips. The favorable test results encouraged the involved parties to test an overloaded Go 242 with take-off rockets. Either two Walter Rl 202b, with 500 kp of thrust each and a burning time of 30 seconds, or four Rheinmetall Rl 502, each with 500 kp of thrust and six seconds of burning time, were used. The Rheinmetall powder rockets were regarded as more practical on the basis of testing and introduced.

Above and below: The principle of the twin-boom transport with folding rear fuselage was adopted by the Americans for their big Fairchild C-82 and C-119 military transports. The Go 242 was a high-wing craft of mixed construction: steel-tube fuselage with linen covering and wooden wings.

This Go 242 bears a camouflage paint job applied by the troops in the winter of 1941-42. The original two-tone green upper paint was painted over with white stripes and spots.

The characteristic double-boom silhouette of the Go 242 soon became known to the army units in the Mediterranean and Balkan areas, and particularly on the eastern front. It was always greeted with relief, especially by the units surrounded at Cholm, Demjansk, etc.

The shark-nose (left) painted on this Go 242 in the Italian-North African zone has an old tradition, for a Roland "Walfisch" (Whale) on the western front bore this design in 1915.

The emblem of Towing Group 4, equipped with Go 242's, indicates something of the danger of their missions.

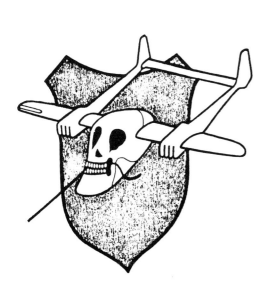

Since no large-scale air-land operations were carried out after the conquest of Crete, the Go 242 was used almost exclusively for supply service afterward. In this capacity, though, it saw service on all fronts. These freight gliders were used successfully even in the African theater of war. On the eastern front the Go 242 became indispensable when it was necessary to supply the fighting army units, especially cut-off groups, since it needed, as already noted, only some 250 meters to land. When the advance in the direction of the Caucasus began in the southeast sector of the eastern front, towing units with He 111 H-6 and Go 242 were used successfully for the first time. In this action, Towing Group 4 (SG 4) stood out particularly. The interior equipment of the Go 242 was changed to suit special purposes. There were versions with built-in bunks to transport wounded men, or flying workshops that came to help the war-zone yards and workshops at the field air bases with spare parts and equipment when necessary.

Above: A Ju 52 tows a Go 242 into the pocket of Demjansk. The pocket at Cholm was also supplied by freight gliders.

Below: At the former Estonian airfield of Dorpat (Tartu), troop testing of the Go 242 on snow skids took place in the winter of 1941-42.

Left: One of the two side snow skids of the Go 242 freight glider. The building of a Go 242 with a floating hull bottom and side pontoons was planned. Whether it was actually built cannot be ascertained.

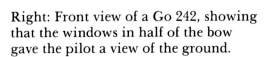

Right: Front view of a Go 242, showing that the windows in half of the bow gave the pilot a view of the ground.

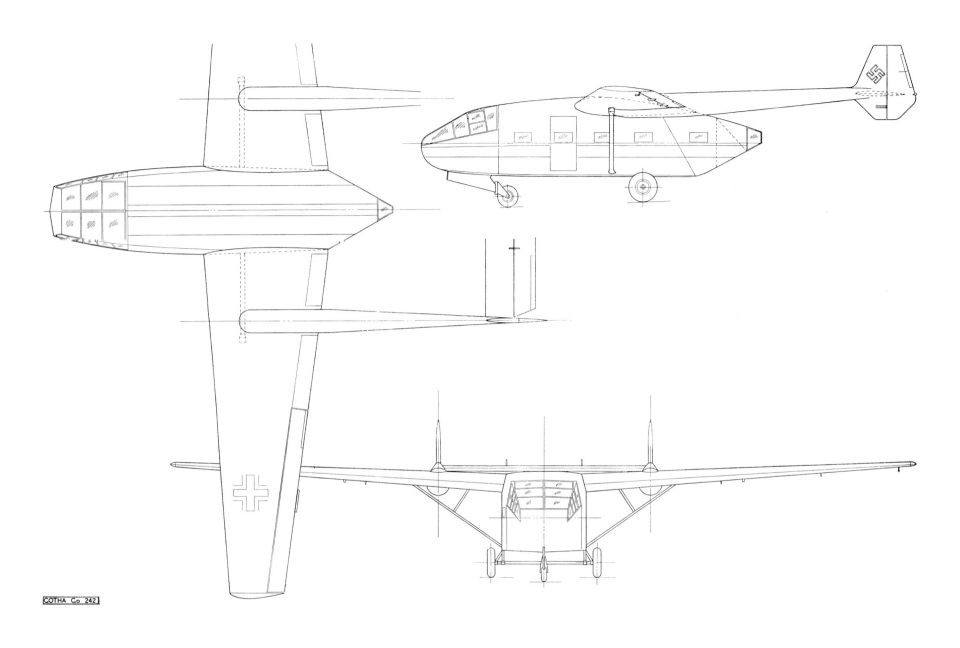

Three-way drawings of the Gotha Go 242 B-1, which already had a nose wheel, later retained in the Gotha Go 244 transport.

To guarantee a sure start, even with an overload, a rack to hold four 109-502 (Rheinmetall R I-502) take-off rockets was built onto the rear fuselage of Go 242 A-1. These solid-fuel rockets weighed 48 kilograms, were 17.8 cm in diameter and 1.27 meters long.

Left: A view of the pilot's seat in the Go 242 with take-off assistance shows how complicated the pilot's work had become compared to the DFS 230.

Right: The pilot's seat of Go 242 B with ignition buttons for the take-off rockets.

Above: A Go 242 takes off with R I-502 take-off rockets, which were ignited at the moment towing began.
Below: The Go 242 has already lifted off and the HWK 109-500 A rockets have ignited.

Above: In cooperation with the Heinkel Aircraft Works, Prof. Walter developed the liquid-fuel (HWK 109-500A) R I-203 rocket engine at Kiel. Among the troops it was known as the "R Device." Here is a Go 242 A with the "R Device."

Left: The "R Device" was dropped after take-off. As it fell, the parachute visible at the head of it unfolded and brought the rocket engine safely to the ground.

In November of 1943, when the 17th Army under General Jänecke was cut off in Crimea, the army was supplied from the Odessa area, chiefly by air across the Black Sea, and the Go 242 played an important role. When Odessa had to be evacuated on April 10, 1944, the Konstanza area became the take-off point for the aerial supplying of Crimea. When Crimea had to be evacuated early in May, the Luftwaffe alone transported 21,457 men, a large number of whom were carried by the He 111-Go 242 towing groups. Other surrounded German support points such as Cholm, Tarnopol, Welikije Luki and, in 1944, Budapest were effectively supplied by these towing groups. It should also be noted that as early as 1942 the Go 242 freight glider had already been transformed into a very useful transport plane, designated Go 244, by installing air-cooled radial engines, generally captured French or Russian units. This development was repeated with the large Me 321 freight glider.

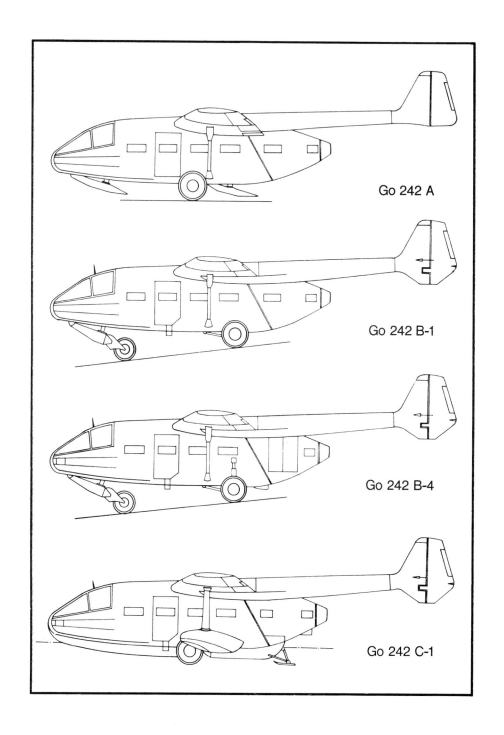

Right: Side views of the various versions of the Go 242.

Left: Since Chief Designer Dipl.Ing. Kalkert wanted to equip the Go 345 combat glider as well as the Ka 430 freight glider with a simple empennage, he had a Go 242 remodeled appropriately. A Horch personnel car is just being driven aboard.

Right: The rear view of this experimental Go 242 shows clearly the loading hatch, which also served as an entrance ramp.

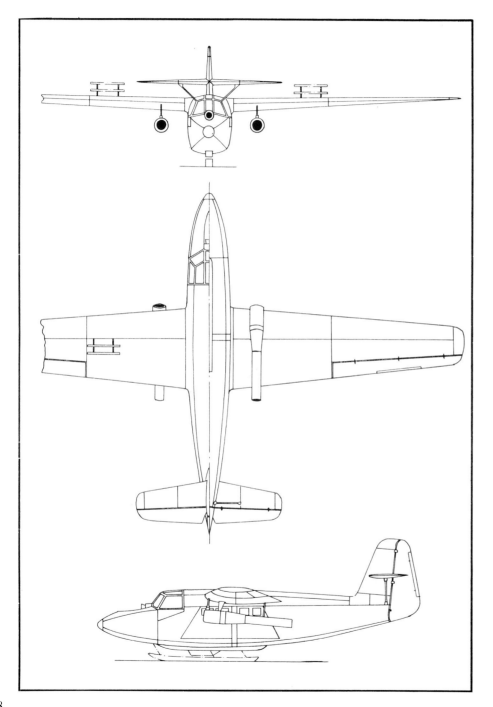

Right page:
Above: The first Messerschmitt Me 321 was towed out of the hall about March 20, 1941. It was not long before series production was going on at full speed in Leipheim. Soon several Me 321's were ready for their first flight.

Below: In the first test flights of Me 321 V-1, Ju 90 V-7, which can be called the direct ancestor of the Ju 290, served as the towplane. This picture shows one of the first take-offs at Leipheim. The wing-tips of the Me 321 can be seen behind Ju 90 V-7.

Left: These three-way drawings of the Go 345 combat glider clearly shows the Argus jet tubes used for take-off assistance, as well as the air brake to decrease the landing speed. There was also a braking rocket in the nose.

Below: Three-way drawings of the Ka 430. It too was built only in small numbers.

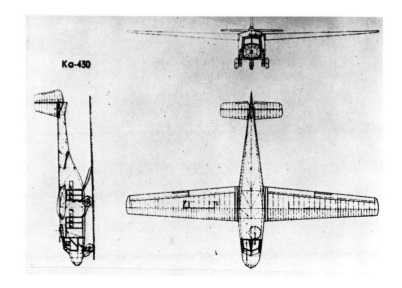

Me 321

When it was seen after the victory in the west in 1940 that the navy was not in a position to provide enough shipping and securing for the intended invasion of England, the operation "Sealion", the idea of landing troops and supplies, including heavy weapons, by air was suggested. At the Reich Air Ministry it was believed that it must be possible to build freight gliders for this purpose, though according to present-day thinking they would have had to be large craft indeed. GL/C, the Technical Office of the Luftwaffe, sent out a description of a large freight glider that would be capable of carrying a Panzer IV tank, an Sturmgeschütz III, a tractor with an 88mm Flak 18/36 plus ammunition and fuel, or 200 infantrymen with full equipment. The program began under the code name of "Warsaw." It was developed by Junkers as "Warsaw-East" and by Messerschmitt as "Warsaw-South." In the process, though, the RLM gave instructions that made Junkers' work difficult from the start, if not actually dooming it to failure: Junkers, the factory that had practiced and developed the construction of metal aircraft from the beginning, was to build a freight glider completely of wood. Neither the designers nor the builders had the least experience for this task. But the Chief of Technical Development at Junkers, Professor Hertel, felt he could handle this assignment.

Messerschmitt had it somewhat easier: Its freight glider was to be of mixed construction, with a steel-tube fuselage covered with linen and wooden wings. "Warsaw-East" was given the type number Ju 322, "Warsaw-South" Me 321.

Left: In test flights with Ju 90 V-7, the Me 321 showed good flying characteristics. But despite servo-assisted controls, the enormous steering pressures made great demands on the pilots.

Right: Rheinmetall R I-203 rockets, two of which were hung beneath each wing, provided take-off assistance. Later Walter HWK 109-500 rockets were also used.

Left: All Me 321's had registrations that began with W. W 1 to W 4, W 6 and W 8 were used by Messerschmitt for their large freight gliders. Likewise, the letters after the cross emblem always began with S. The craft shown here, W6+SW, is a typical example of these registrations.

Right: With a wingspan of 55 meters, the Me 321, and the subsequent Me 323, even exceeded the dimensions of a modern large-capacity airplane like the Boeing B 747. When one flew over at a low altitude, one could feel the pressure.

The Me 321 was to be built at Leipheim, the Je 322 at Merseburg. On November 1, 1940 the design drawings had to have reached the RLM. Just six days later, the contracts to build them arrived by telegram with the comment, "Program to start at once."

Messerschmitt was also told, "Build twice as many. Very urgent." A hectic work tempo began at both Merseburg and Leipheim. Three months later, in February of 1941, Messerschmitt finished the Me 321 V-1 model. The landing gear could be dropped, and the craft could land on skids. The landing gear alone weighed 1700 kilograms! Only the four-engine Ju 90 could serve as a towplane. Eight of the so-called "R Devices", each providing 500 kp of thrust, were planned for take-off assistance. The first take-off took place on February 25, 1941. Despite servo controls, steering the giant glider demanded the pilot's last bit of strength. Along with the factory pilots, Germany's best glider pilots, including such men as Bräutigam, Kraft, Flinsch and Böhm, served as pilots. They got along very well with the Me 321, and their gliding experience was extremely valuable in testing it.

Above: The troika take-off was always a risk for all involved. The take-off did not always go as well as here. The Me 321 has already lifted off. The three Bf 110's carefully pull the giant glider higher.

Right: Compared to the mighty Me 321, the three Bf 110's look quite tiny. Matching the speeds of the three towplanes exactly demanded ability and cool heads of their pilots.

Above: Preparing an Me 321 for take-off in Russia, in the autumn of 1942. This Me 321 already has an enlarged rudder to improve vertical stability. The comparison with the two men of the ground crew shows especially vividly how gigantic the dimensions of the Me 321 were in its day.

Left: The Me 321 stands fully loaded, serviced and ready for its next supply flight.

It proved to be practical to enlarge the cockpit of the Me 321 and install double controls. At Leipheim, production became faster and faster. This raised the question of sufficient towplanes, for only a few Ju 90's were available. Thus Messerschmitt suggested multiple towing. For this purpose, three Bf 110 C-1 planes were to tow one Me 321. This procedure later became known as the "troika tow." With a weight of 20 to 22 tons, such a take-off required 1200 meters of paved runway, naturally with the assistance of R Devices.

It soon became obvious that the troika tow was dangerous. Again and again, accidents happened during take-off, which demanded the highest degree of concentration and precision flying from all involved. In an especially bad case, the entire combination crashed. Only some of the R Devoces ignited, pushing the Me 321 so far off course that the three Bf 110's collided. Nine men of the flight crews and 120 infantrymen aboard the Me 321 lost their lives. In the summer of 1941, the Me 321 was accepted by the Luftwaffe and taken to France. There were still hopes of carrying out Operation "Sealion." When this was finally given up, Operation "Barbarossa" had become perilous. It was decided to make a virtue of necessity and set up three Me 321 units for supply purposes. In the spring of 1942, four more units were sent to the northern part of the eastern front, in the Riga-Orscha area.

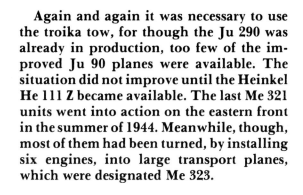

Again and again it was necessary to use the troika tow, for though the Ju 290 was already in production, too few of the improved Ju 90 planes were available. The situation did not improve until the Heinkel He 111 Z became available. The last Me 321 units went into action on the eastern front in the summer of 1944. Meanwhile, though, most of them had been turned, by installing six engines, into large transport planes, which were designated Me 323.

Upper left: The four R I-203 solid-fuel rockets are giving full thrust, and the giant craft is gliding safely and quietly upward.

Center: The towline from one Bf 110 C broke, the plane crashed and burned on impact. The other two planes immediately released.

Bottom: The pilots of the Me 321 were able to land the craft near the field without major damage. The W 2 registration shows that this Me 321 came from the second production run at the Messerschmitt factory in Leipheim.

Upper left: On the initiative of Generalluftzeugmeister Udet, who rejected the troika tow as too dangerous, the Heinkel He 111 Z took shape at Rostock-Marienehe. This is one of the first two models in the autumn of 1941.

Left: The He 111 Z had five Jumo 211 F-2 engines of 1340 HP each. This is the central motor. In the background is the Me 321 to be towed.

Above: According to the pilots who flew the He 111 Z, this plane, which was only a makeshift solution, was fairly easy to fly despite its ponderous handling and high steering pressure in curves.

In 1943 the He 111 Z-Me 321 combinations were used mainly to supply the Kuban bridgehead. The supply depot was Bagerowo, in Crimea. The targets were Krasnodar, Timashevskaya and Slavyanskaya. On the return flights, only wounded men were carried. The Me 321 carried up to a hundred men, while the He 111 Z, which was not at all planned for this purpose, carried another thirty.

Ju 322

The Ju 322 proved to be a failure in every respect. When the wood for 100 craft was delivered, it turned out that it was moldy. The first wing spar, which was supposed to carry a 1.8-fold load, broke with a load of 0.9, the second at 1.1! The droppable starting cart bounced so high after being dropped that it endangered the glider. The craft weighed 8000 kilograms. When a Panzer III tank rolled up the too-steep ramp into the cargo hold in the first loading tests, it tipped over forward and broke through the floor. Instead of the required 20-ton payload, it could carry only 12 tons. Then too, the defenses at the Junkers plant had worked so badly that in March of 1941 BBC-London was already able to give details of the "Merseburg Giants." A few days before the first take-off, Generalluftzeugmeister Ernst Udet inspected the Ju 322, which was also called "Mammut" (Mammoth)! His reaction was depressing. He said what many experts had already suspected. "This aircraft will never fly. The proportions of the empennage are wrong. The machine will never be stable!"

The first flight proved that his evaluation was correct. The Ju 90 rolled away, connected to the Ju 322 by a 120-meter towing cable (16 mm diameter). Despite full throttle, the combination never reached the speed needed to lift off. Only shortly before the end of the runway did the Ju 322 get off the ground. The impact smashed the dropped landing gear. The craft was so unstable that it almost made the towing Ju 90 crash. The pilots quickly disconnected the towing cable and tried to bring their craft to the ground in a flat glide. It was impossible for them to correct its course. Thus the landing took place far from the airfield. It was very difficult to bring the giant back to the airfield and rebuild it. The empennage in particular was redesigned.

It is not known whether the plane ever got off the ground again. In any case, the entire program was halted. Material had been provided for 100 gliders, of which thirty were under construction. All the parts were sawn up and the pieces used in the wood-gas generators often used on motor vehicles at the time. The whole undertaking had cost 45 million Reichsmark, quite apart from the wasted work. But it was not the only project for which materials and workers were used in vain!

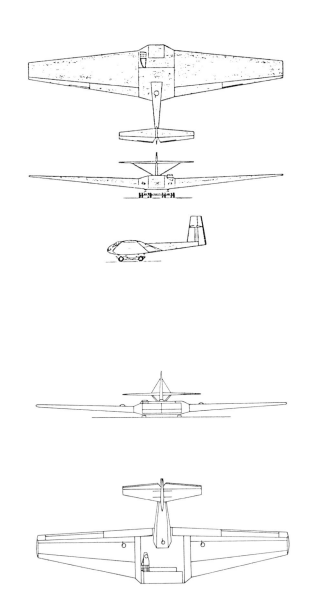

Upper right: Early form.

Right: Final form of the Junkers Ju 322 V-1. Professor Hertel's original design for the "Mammut" glider bore design number EF 94.

Technical Data

Craft	DFS 230 A-1	DFS 230 V-7	DFS 331 V-1	Go 242 A-1	Go 345	Ka 430	Me 321	Ju 322
Crew	1	2	2	2	2	2	2	6
Wingspan (m)	21.98	19.40	23.00	24.50	21.00	19.50	55.00	82.35
Length (m)	11.24	12.50	15.81	15.80	13.00	13.22	28.15	–
Height (m)	2.74	2.90	3.55	4.26	4.20	4.17	–	–
Wing surface (sq.m)	41.30	39.50	60.00	64.40	49.90	39.90	300.00	–
Empty weight (kg)	860	1253	2270	3200	2470	1810	12000	–
Payload	8-10 men	15 men	?	3564-4064 kg	10 men	12 men	–	140 men
Flying weight (kg)	2100	2400	4770	6800-7300	6000	4600	34000	–
Max. towing speed (kph)	290	320	330	240	370	320	218	–
Cruising speed (kph)	210	300	270	210	310	300	200	–
Ceiling	1600	?	?	1800	?	–	–	–
Two-way range (km)	?	?	?	375	?	–	–	–
One-way range	?	?	?	670	?	–	–	–
Armament	1 MG 15	–	2 MG 15	to 8 MG 15	?	1 MG 131	4 MG 15	3 MG 15

Schiffer Military History

Schiffer Publishing Ltd.
1469 Morstein Rd.
West Chester, PA 19380
Phone: (215) 696-1001   FAX: (215) 344-9765

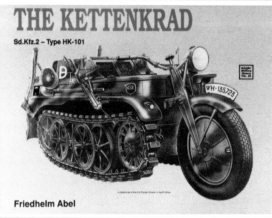

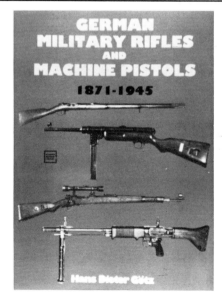
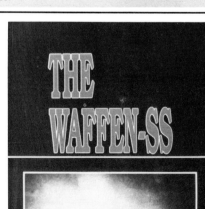